Edward Santiago

THE ROUND

An Introduction

A facsimile of the original 1942 edition

Edward Santiago

THE ROUND

An Introduction

A facsimile of the original 1942 edition

*With an introduction by Judith Hipskind Collins
and a preface by Paul Herron*

Sky Blue Press
State College, PA

Copyright © 1942 by Edward Santiago

All rights reserved. No part of this book may be reproduced in any form or by any electronic or mechanical means, including information storage and retrieval systems, without permission in writing from the publisher, except by a reviewer or scholar who may quote brief passages in a review or article. Facsimile version copyright © 2025 by Sky Blue Press.

Introduction copyright © 2025 by Judith Hipskind Collins

Published by Sky Blue Press State College, PA
www.sky-blue-press.com

Library of Congress Cataloging-in-Publication Data

Names: Santiago, Edward, author.
Title: The round : an introduction / by Edward Santiago.
Description: State College, PA, USA : Sky Blue Press, [2025] |

Identifiers: LCCN 2025021998 (print) | LCCN 2025021999 (ebook) |
ISBN 9798985524086 (hardback) |
ISBN 9798985524024 (paperback) |
ISBN 9798985524093 (kindle edition)
Subjects: LCSH: Occultism. | Astrology.
Classification: LCC BF1999 .S33 2025 (print) | LCC BF1999 (ebook) |
DDC 133.5--dc23/eng/20250715
LC record available at https://lccn.loc.gov/2025021998
LC ebook record available at https://lccn.loc.gov/2025021999

PREFACE

THIS EDITION OF *The Round* is a facsimile of the original 1942 book, which was printed in a 125-copy run. The copy used for this rendering was once owned by Martha Jaeger, a psychoanalyst whose patient was Anaïs Nin, Eduardo Sánchez's beloved cousin. The only alteration was to slightly enlarge the text for the sake of readability. In these pages, there are mentions of Nin and some of her associates, including Otto Rank, Antonin Artaud, and René Allendy, whom Sánchez calls "Dr. X."
—*Paul Herron, April 2025*

INTRODUCTION

STARGAZER EDUARDO SÁNCHEZ (*nom de plume* Edward Santiago) was born in Cuba in 1904. He was the third of seven children and grew up in a family who owned a portion of the country's sugar cane industry, as well as other business interests. While most members of his family were boisterous and outgoing, he was introspective and sensitive, a dreamer who loved solitude.

At age six, he witnessed the appearance of Halley's Comet. That event made an indelible impression, and when he learned the Comet would return years later, his fascination with celestial cycles was born. He was determined to live to see the Comet's return.

When he was twelve years old, he was sent to a boarding school in the Hudson River Valley, in New York. There he excelled in his studies, navigating in English for the first time. He developed an ear for languages, ultimately also becoming fluent in French, Italian and German. Along the way, he mastered Latin, using it in his written research. This ability often gave him access to information at its original source, contributing greatly to his quest for knowledge.

His college years were spent at Harvard. While there, he discovered acting and the theater, and threw himself wholeheartedly into this world. He became President of the Hasty Pudding Club. After graduation, his passion for acting led him to Broadway where he appeared in several plays. He received good reviews. But in his short time on Broadway, he never had the opportunity for a leading role that he wanted. Although he was ambitious, drawn by his passion for acting, he began to think his chances for a leading role were hampered by his Spanish accent. He imagined he would be limited to character roles. He grew discouraged and returned home to Cuba.

Shortly after going home, he fell ill with an infection following a minor surgery. The infection quickly became dangerous, and only the arrival of an antibiotic from Miami saved his life. This was the earliest version of newly discovered penicillin which his family was able to get.

After months of recovery from this strange event, he came to a decision, made together with his father, that he would not work in the family business. The alternative was to leave Cuba and go to Europe where he would travel and pursue his interest in art, literature and philosophy. He wanted to meet the literary figures he admired, and his biggest desire was to meet writer D. H. Lawrence.

He began the most important years of his life, ultimately involved in travel and research, in August of 1930. When he arrived in France, hoping to meet Lawrence, he learned that Lawrence had died a few months earlier, in March 1930.

Eduardo went to Paris where he reunited with his cousin Anaïs Nin, who was living there with her husband Hugh Guiler. In Paris he found a whole new world of creativity and camaraderie. He joined writers, artists and thinkers in a circle that included Anais, Henry Miller, and other names to be found in *The Diary of Anaïs Nin*.

While Henry Miller painted and wrote, Eduardo and others in the group became enchanted with the work of W. B. Yeats. They adopted *A Vision* as their own, writing their responses to Yeats' words in the margins of a copy of the book. Occasionally Henry added his thoughts to those pages as well.

Enthusiasm for the work of Yeats inspired Eduardo to follow the thread of astrology. Here he found his childhood theme of cyclical returns of the heavens' phenomena that had fascinated him about Halley's Comet. All the planets moved around the Sun in differing time frames, but always returning to their starting point in the Zodiac sign of Aries. He was an Aries himself with the planet Jupiter in Aries. In astrological interpretation, the Sun symbolically represents a person's identity. Having Jupiter very close to the Sun on the day of his birth encouraged him to view his life path taking Jupiter's position into consideration. Jupiter's orbit around the Sun takes 12 years. And given his age, Eduardo had just completed three of those 12 Jupiter year cycles when he began work on *The Round*. What an opportunity to explore his life framed in those 12-year segments. Filled with inspiration, he began *The Round* as his Magnus Opus, bringing together his experiences of heightened identification with the sources of what propelled him along his path to discover not only himself but to understand the very nature of reality. He describes his search, revealing his experiences, his emotions, and yearning on the path to a wider sense of a purpose-filled self. He does this with grace, elegance and persistence.

His search involves a wide range of philosophical references, eras of history and famous figures, along with a more in-depth explanation of planetary correspondences accompanying them.

In *The Round* can be found the seeds of his lifelong journey with astrology. In Chapter XII, he defines his goal of examining and isolating the trait of genius in famous people with substantial accomplishments in a wide range of fields. He continued this investigation long after completing *The Round*.

Traveling throughout Europe, he compiled the birth data needed to set up the birth charts of hundreds and hundreds of famous people. He was always looking for that elusive germ of genius as defined by the planets in each individual chart. In his work he studied every type of writer, artist, scientist, philosopher, inventor, famous Renaissance figures, clergy including long lines of Cardinals and Popes. He was especially intrigued with the Medici family, studying them, their charts and their influence in the world.

His work in astrology took one final direction. When he arrived in Europe in 1930, Pluto had just been discovered. The addition of a new planet whose orbit around the Sun was the longest yet discovered inspired him to consider it along with the two closest neighboring planets, Uranus and Neptune. He correlated the cycles of these three planets taken together to examine known periods of human history.

Of course he found the cycles and the patterns of their return, taken together in their various angles, corresponded with much of known history. From that he projected his thoughts for epochs yet to come. He had moved from the study of genius as seen in the patterns of the faster moving planets to a truly cosmic view of history's potential.

For a young boy who once fell in love with Halley's Comet, he had come a long way. And his wish was fulfilled. He did live to see the Comet's return in 1986.
—*Judith Hipskind Collins, April 2025*

THE ROUND

THE ROUND

AN INTRODUCTION

EDWARD SANTIAGO

MORNING STAR FARM
ASHFIELD, MASS

COPYRIGHT, 1942
BY EDWARD SANTIAGO

THE INTRODUCTION:

APOCASTASIS
AND
THE EGO

ILLUSTRATIONS

A 'KING OF JUDAH', ON THE WEST PORTAL OF THE CATHEDRAL AT CHARTRES: FRONTISPIECE

ST. SEBASTIAN, BY PERUGINO, AT ROME   Page 16
[Through the courtesy of the Metropolitan Museum of Art]

ST. CHRISTOPHER, BY POLLAIUOLO   Page 25
[Through the courtesy of the Metropolitan Museum of Art]

LIFE MASK OF WILLIAM BLAKE, in the National Portrait Gallery, London   Page 40

THE FIRST DATED WOODCUT OF ST. CHRISTOPHER, 1423   End Pages

ST. SEBASTIAN [Taken from 'Pest Blaetter der XV Jahrhunderts', edited by Paul Heitz]   End Pages

CONTENTS

CHAPTER		PAGE
I.:	PALINGENESIS	1
II.:	RELIGIO MEDICI	2
III.:	CIRCUMAMBULATIO	4
IV.:	AB OVO	9
V.:	HERCULES FOURENS	10
VI.:	PSYCHOANALYSIS	13
VII.:	COINCIDENTIA OPPOSITORUM	14
VIII.:	HEAUTONTIMORUMENOS	17
IX.:	HERCULES OETAEUS	18
X.:	DEUS EST DEMON INVERSUS	21
XI.:	COMMUNIO SANCTORUM	25
XII.a.:	ASTROLOGIA REDIVIVUS	31
XII.b.:	ASTROLOGIA LITERARIA	33
XIII.:	JOURNEY OF THE MAGI	41
	EPILOGUE	46

'If in an experience one does not stake his reason, that experience is not worth the attempt.' Gaston Bachelard: Le Surrationalism, from INQUISITION, June 1936.

I. *PALINGENESIS*

IN these livid decades of Psychoanalysis and the Objective Hazard of Surrealism, it is not surprising that I am eager to trace within myself the genesis of an idea, rather than to give an objective, or even an artistic, presentation of that idea.

Dead to us is the 'Discours de la Methode pour bien conduire la raison et chercher la verite dans les Sciences'. But with what excitement could we not have turned to the 'Olympica' if it had not been permitted to be lost. For in that work, Descartes had described the outbursts of the Irrational in those few memorable November nights which led to the synthesis of geometry and algebra.

But my purpose here is not only to trace the genesis but also the palingenesis of an idea, an idea

APOCASTASIS

that once had reached such a rigidity of dogmatic pronouncement that it lost all reason for subsistence; an idea that reissued, in all its completeness, in the mind of the most provoking genius of the nineteenth century: Nietzsche, who had searched for the most difficult belief to which he could give allegiance, and found in the conception of Eternal Recurrence the incitement to a joyful consent, even affirmation.

And, once more, my purpose is not to describe an identical performance, but to retell the rebirth and show the transformation of the Idea undergone during its regestation. What had once been an intuition, and then a dogma, has now become, by its application to the progressed data of modern astronomy, a workable, an empirical hypothesis. Only the future can verify, retransform, or reject it.

II. *RELIGIO MEDICI*

Already, to my knowledge, a book has been published, introducing the imprint of the moving Macrocosm on the living microcosm, and giving the necessary perspective to a work which presents the opinions of a doctor on various fundamental problems: 'Religio Medici', 1643.

APOCASTASIS

This book was written at the moment when Descartes was emancipating the reason from all contingencies, isolating Man from the cosmic environment and consequently giving the decisive impetus to the Promethean movement of Scientific Progress; and at the time when Locke was supposedly collecting the sense-data which were later to converge and be transformed into the idea that time was not circular but straight.

In this book Sir Thomas Brown informs us that one revolution of Saturn [29 years and six months] had not yet taken place in the life of the writer. And, further on, he reveals the two most salient points in his horoscope.[1]

But whereas Sir Thomas' outlook was adequate to the age in which he lived [the great Newton, who was later to play so creatively by the Seashore, was but a year old], my purpose in introducing myself in this present work transcends the intellectual

---

1. 'If there be any truth in Astrology, I may outlive a jubilee; as yet I have not seen one revolution of Saturn, nor hath my pulse beat thirty years.'... Page 47.

'At my nativity my Ascendant was the watery sign of Scorpio; I was born in the planetary hour of Saturn, and I think I have a piece of that leaden planet in me.'... Page 85.

RELIGIO MEDICI, New Universal Library, George Routledge and Sons, London.

APOCASTASIS

honesty to present my personality as a prism through which the 'eternal' problems are to sift, acquiescing to the dictum: My world as a representation.

I present myself as a point of interference for the inward and outward dynamic processes which culminated in the rebirth of Apocastasis, or The Great Year.[1]

III. CIRCUMAMBULATIO

With my escape from Europe in June of this year of 1940, there ended a physical journey begun in 1928.

If we were to trace the itinerary of this journey on a map of the Northern Hemisphere, we should find a curving line that starts at New York, passes through the Eastern States, and reaches down to my home in the West Indies. It lingers there for a fat year and a lean year, and then rises up to Europe by way of Spain. Finally, after ten vindemial years, it

---

[1]. 'Apocastasis is the word which the Chaldeans had already used to describe the return of the planets, on the celestial sphere, to the points symmetrical to their departure. It is also the word the Greek doctors employed to describe the return of the patient to health.' J. Carcopino, VIRGILE, Paris. 1930.

APOCASTASIS

takes to the perilous waters of the Atlantic, zigzagging from Liverpool to Montreal.

And I am back in New York.

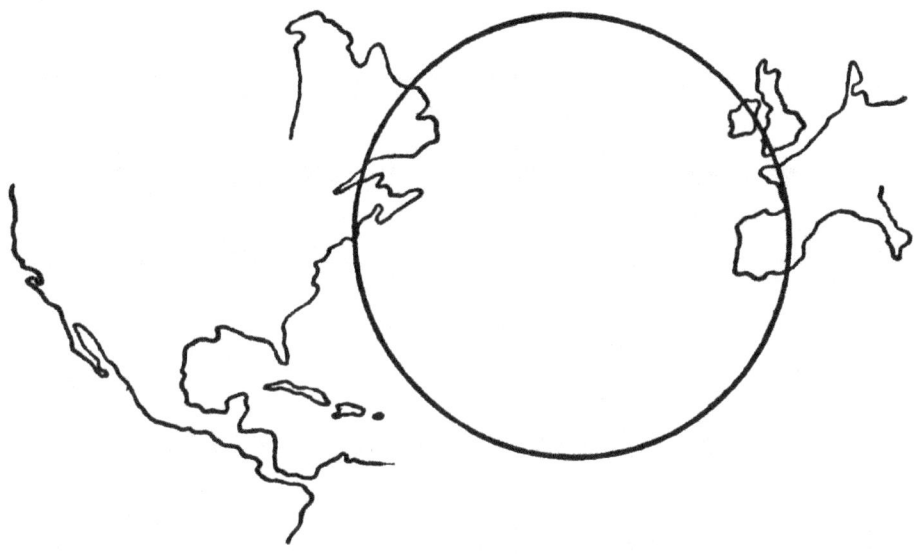

Thus, we can see that in a span of twelve years I have followed the circumference of a circle in Space.

But it is as Time that we need to consider this circle, for Time, unlike Space, still contains an element of the 'miraculous'. The exploration and the conquest of Space give us no hesitancy. But Time confronts us as an enigma. It is only in quite recent decades that by the welding of these two concepts,

APOCASTASIS

and therefore returning to the conception previously held by the Ancients, that we begin to understand that the enigma was only due to a confusion of thought, or rather to the sundering of concepts.

For to the Ancients, Time was not a pure abstraction, but the obvious equivalent of the planetary movements. There was no Absolute Time, but many times, all depending on the small or large cycles of the planets, and the earth and every being on it was subjected to such times. In short, the Ancients, after their own fashion, had a true conception of our 'Time-Space-Continuum', a truer conception, since they founded on it a philosophy of history still worthy of our attention. But we shall not be able to appreciate its application to our modern problems until we firmly repudiate the Physicist's depreciation of astronomic time as time-keeper [proposing in its place the time-span of a disintegrating atom of radium]; or Bergson's contemptuous treatment of it, preferring, in his premature idealism, the continuum of subjective time.[1]

---

1. Premature because it is only after a thorough study of astronomic, or objective, time and the strength of its determinism, that we shall be able to approach with any hope of success the problem of subjective time and its relation, if any, to astronomic time; and success will only be forthcoming when both concepts are given equal validity.

APOCASTASIS

Let us then apply this conception of Time to the circle which I have traced on the Atlantic corner of our earth within the span of twelve years.

An investigation into the planetary cycles will show that this span of time synchronizes with the twelve-year revolution of Jupiter around the earth.

I was born in 1904 when the Sun (☉) was in conjunction with Jupiter (♃) in Aries. When, after a lapse of twelve years, Jupiter returns to the conjunction, I leave the West Indies for the United States. Precisely twelve years later, as the same astronomical event is repeated, I find myself returning to the West Indies.

Now another cycle has been completed. It is 1940: and I am back, a fugitive from a convulsed Europe.

It is obvious that this circular trajectory in time had its counterpart in space. And it is obvious that this rare concurrence could not have taken place during the two cycles comprising the first twenty-four years of my life, since in that period of immaturity I was subjected to more decided wills than my own. It was only after I had left college that I was completely 'disposable', that is, free to follow my deepest needs and the tempting stimuli of the

APOCASTASIS

world. In this 'susceptible' condition, Jupiter became insistent and, by his silent impulsion, the synchronization was accomplished.

A reference to this Jupiterian cycle is to be found in Censorinus' 'De Die Natalie', 238 A.D.:

The most exact Great Year is the dodecaeteries, which is composed of twelve solar years. It is called the Chaldean Cycle*. The Astrologers did not regulate it by the course of the sun or moon, but after other observations, because they said that only this space of time could embrace the different seasons, the epochs of abundance, of sterility and of plagues ('tempestates, frugumque proventus, sterilitates item, morbosque circumire').

At this juncture, anyone familiar with Plato's philosophy of history should not fail to recall that passage in the REPUBLIC [8:56]:

That a city so constituted should change is difficult; but, since decay is the lot of everything that has come into being, even this constitution will not abide forever, but will be dissolved. And its dissolution will be as follows: To all living things, and not only to plants that grow on the earth, but also to animals that live upon its surface, come time of fertility or barrenness of soul and body as often as their revolutions complete for each species the circumferences of circles, which are short for short-lived, and long for long-lived...

* William Maude, the American translator of this work for the year 1900, calls it the Jovian Cycle.

Can there be a direct, or indirect, connection between these two passages, formed by the tradition that there was a Chaldean among Plato's disciples, and the knowledge that the enthusiasm of Eudoxus for the astronomic data of the East spurred Plato's curiosity?

Be that as it may, I take over the Chaldean Cycle; and I will show how, in my time-space circuit of twelve years, I also have undergone seasons of abundance, sterility, and even of 'morbidity' or 'neurasthenia', to use an old-fashioned term in Psychiatry.

IV. AB OVO

Whether 'Creation' has had a beginning and, therefore, will have an ultimate end, I leave for the future to decide. But there is no question that in world history there are intermediary beginnings and intermediary ends.

If we apply a schema of closed cycles, it would facilitate our search for the demarcations of such beginnings and ends, for in the Cycle we meet the two great bugbears of Physics: the continuous and the discrete. In fact, the Cycle is a circle of continuous discreteness [the cinematic of angles], wherein the end-beginning is its major nodal point.

APOCASTASIS

For example, my departure from New York in 1928 was the end of a cycle and the beginning of another. Here we have the starting point of my 1928-1940-Experience. But if the import of this Experience is to be understood, I find it necessary to relate at least the last phase of the previous cycle, in order to reveal how the end became the beginning.

The Sine-Wave of the Chaldean Cycle

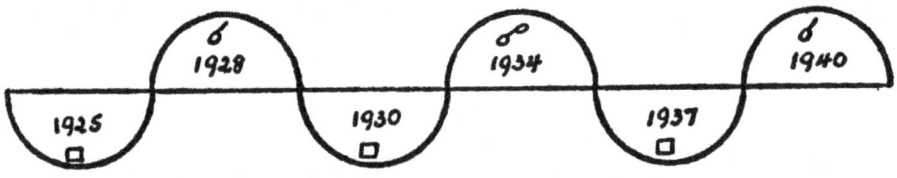

V. HERCULES FOURENS

Without giving any psychoanalytical reasons for my attitudes, I shall straightway state that I have been periodically impelled by grandiose ideals. At such periods I have not been able to prevent these ideals from inflating too far beyond my immediate capacities. Accordingly the premature and impatient attempts to actualize them would hurl me into a corresponding state of deflation, which was

APOCASTASIS

unbearable in its weight of revulsion, self-accusations and schizophrenia.

This can be illustrated by the episode of my life comprised in the last phase of the 1916-1940 Cycle:

The predominant activity of this cycle was the Theatre. I had been president of the dramatic club at high school and at college, and when I came to New York, it was to become, not an actor, but The Actor.

I set about it in my over-ambitious manner:

I read and re-read Keyserling's 'Travel Diary of a Philosopher', that passionate attempt to lose one's personality by the arduous exercise in understanding of the various cultural races of the East; an exercise, I was firmly convinced, which would give me, The Actor, the metaphysical foundation which Keats demanded of the true poet.

And to give structure and a time-perspective to these and other cultures, I studied that astonishing tour de force, 'The Decline of the West', by Oswald Spengler.

As for the passions, the normal, the profound and the perverse, I turned to Dostoievsky, Blake, Whitman and D. H. Lawrence, and soon Titans were whirling in the enormous stage of my imagination, as set by Blake and Edward Craig.

APOCASTASIS

But I was to understand also the delicacy, the refinement, the finesse of psychological processes at the microscopic hands of Marcel Proust, and I immediately connected the 'Intermittances du Coeur' with the remarkable technique then being taught at the American Laboratory Theatre in New York by Richard Boleslavsky and the great Oupenskya.

The outcome was, of course, a split between the Ideal and the Actual. The more certain I was of having achieved the death of my personality, and therefore the birth of the potentiality of all personalities, the more disappointed I became with my body as a fitting vehicle. Its definiteness of form, its slight margin of pliability disgusted me. No amount of physical training, no amount of reading of other actors' achievements [for example: that Richard Mansfield could add height to his figure by facing the audience at a certain angle; or that Salvini could climax a scene with an horrowing shriek which his opened mouth never uttered, and yet the hypnotized tympanums of his admirers had rung with the fearful vibrations], could persuade me of the reconciliability between my body and the Ideal. I was beyond persuasion.

And an epoch of morbidity followed.

APOCASTASIS

VI. PSYCHOANALYSIS

I had reached such a point of dissociation that I could no longer help myself.

I was analyzed by a pupil of Otto Rank.

Slowly the Theatre became mirage and then completely vanished. In its place there was to be found nothing that promised an immediate fulfillment,--nothing but the imperative need to return home in the role of a prodigal son. To this need I finally surrendered, in spite of the analyst assuring me that if I did return the 'cure' would prove a miscarriage.

Thus, leaving New York for the West Indies, I closed the 1916-1928 cycle. Had I known all this, my future may have been different.[1] In the last instance, I would certainly have joined with the analyst in questioning the wisdom of my desire to return, not only because it indicated my refusal to achieve independence, but also because I would have had to realize that what I felt was an end was, at least, the

---

1. It will require a chapter to do justice to this extremely important problem. It is a grave concern, not only to the short-lived individual in relation to the short cycles, but also to Societies, subjected to cycles of long duration. I shall deal with it when the time comes to discuss our modern era.

APOCASTASIS

beginning of a new attitude towards the Theatre.

Nevertheless, the fact that this point in time, promising a fresh impetus of activity, failed to be a trump-card in the hands of the analyst; and the certitude I now hold, as I stand at the end of the third cycle embracing these two spheres of Experience, of the wisdom of that return [reculer pour mieux sauter], are proofs that the attempt to become The Actor was the essential preliminary to become The Spectator in the next cycle.

For the problem underlying this last phase had been one of Understanding. It was only the primitiveness of my psychology, my inability to dissociate the function of cognition from the instinct of imitation [an association which Levy-Bruhl has named 'la participation mystique'] that made me persist along these gradually diverging track-lines, until my sense of equilibrium finally broke down.

VII. *COINCIDENTIA OPPOSITORUM*

With my arrival in the semi-tropics of the West Indies, there began the first phase of my new lease of life. It was none other than of exploring the Actual in the varied play of emotions and instincts;

APOCASTASIS

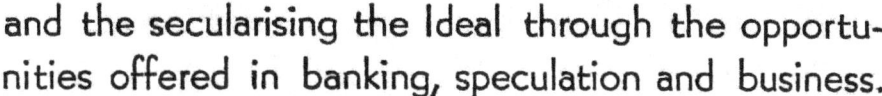

and the secularising the Ideal through the opportunities offered in banking, speculation and business.

But what seemed to be a successful 'adjustment' was only a blind to cover the revenge of the Actual. A superficially social conformism balanced the fear of overindulgence, leading to a guiltless exploration of living.

It was good to be alive, not because of any precise historical revolution, but because life in the West Indies was expansive, prodigal, intense, crude, primitive and perverse.

No sooner did I give myself to this whirlpool of sensations and emotions, than I reached the depths where, in the terrifying alchemy of physiological conversions, pain becomes pleasure, and pleasure, pain.

It was only later, when books took hold of me again, that re-reading Blake's 'Marriage of Heaven and Hell', I understood its genius, and realized that there could be no valid metaphysical system without the 'contraries' as a foundation. And I shall never be able to describe the enthusiasm I experienced in the last phase of this cycle, when I discovered that daemonic metaphysician of the Renaissance, Nicholas of Cusa, who, in creating the synthesis of past and present epochs, conceived God-the-Absolute-and-

APOCASTASIS

Infinite as the Conciliation of Contradictories.

But accompanying this abandonment was the slow-growing conviction that my life should be a development in concentric, enlarging circles in space [possibly as a counterbalance to the deepening of my emotional experiences]. That is, issuing from the womb, I had explored the room, the house, the town, the capitol of the Island, then the States: New York. And now it was only natural that the next circle should comprise Europe: Paris, London, Rome, Berlin. Then the East...... Then, at last, the jump in consciousness when I would 'stand' on the parapet of the 'Fixed Stars' and embrace the harmony of the Solar System.

So I spread out maps on the walls of my study, and placed a revolving atlas on my desk, and peered at the heavens in the look-out for the constellated vantage-point,.....waiting for the opportunity to depart across the Atlantic.

But this was a slow-growing conviction. It was not compelling, and the waiting did not become desperate, until the exploration of the Actual appeared to have been exhausted. Life in the West Indies began to pall.....

And an epoch of morbidity set in.

VIII. *HEAUTONTIMORUMENOS*

This time there was no psychoanalysis to help me, and I fell into a slough of defeatism, and consequently into an unconscious determination to self-destruction.

The unconscious will. What a tremendous power it has! The various ways by which it accomplishes its ends leave us in the same astonishment as the ways of Providence in the Old Testament.

But the question arises: Did I 'suggestion' It either to leave the West Indies and return to the Theatre in the States [this being the most feasible step to take to achieve independence], or else to destroy me? Or did It give me to undersand that if I did not leave, it would destroy me? Which ever way we may be inclined to answer this question, it would not alter the fact that It did almost achieve my undoing.

I am forced to relate the essentials of this experience, since it had its consequence in the next phase of the cycle.

In this state of morbidity I was persuaded by a doctor friend to be operated for a large mole I had on the center of my chest. He assured me it would

inevitably lead to cancer. But this simple extraction gave way to complications: Septisemia: four major operations, --the last one for a mistaken pleurisy, bringing a foretaste of death for which I was so eager. Fortunately, the quietus failed to come. The parental means and solicitude pooled all the medical resources and defeated my will's purpose.

When I recovered, I decided definitely to act on the other alternative of my will. I began to make plans to leave for the States, and with my usual honesty I divulged these plans to my father. Needless to say, it precipitated a storm. In vivid shots of lightning he categorically refused to let me go.-- In the downpour of rain he relented. Yes, I was in need of change. But I must not return to the States. He would send me to Europe.

Suddenly the rainbow.....
And to Europe I came!

IX. *HERCULES OETAEUS*

Two images juxtapose themselves in my mind: the birch and fir-covered coast-line of Newfounland in the grey, crisp afternoon of June 9th, 1940; and the green, old, soft-rolling hills of the northern shore

APOCASTASIS

of Spain in the warm, sunlit morning of July 1st, 1930. Two contrasted landscapes: one, wild, fresh, unexplored, unconquered; the other, softened and tamed by generations and generations of Europeans. Wild nature; domestic nature. These two images form the limit of my horizon encircling my ten years' stay in Europe.

The emotion associated with the image of the Spanish coast was one of profound elation. With the other I feel the aftermath of horror, suffused with the satisfaction of having escaped, and illuminated by the calm certitude that on this hemisphere a new mutation in consciousness was to emerge at the conjunction of Pluto, Neptune and Uranus about the years 3370 A.D., as the previous mutation had emerged 3940 years before among the Greeks in Ionia. Small consolation, it is true, for the knowledge of the Darkness that is now falling...

But in that moment of my profound elation at the glimpse of the coast of Europe, I felt the sense of relief for having escaped from the Western Hemisphere and the irrevocability of my past....

After some weeks of travel in Spain, France and England, I settled down in a studio by the cemetery of Montparnasse, in Paris; and I turned with eagerness

to the development of a hitherto neglected talent: drawing, a talent that insisted in expressing itself sporadically during my psychoanalysis, as though hinting it could well serve as the pivot on which I would turn, if I wished it, the new chapter of my life.

I was to become The Artist, The Prophet-Artist, one of the links in the Golden Chain of prophet-artists to which William Blake belonged. And, as with William Blake, verse and the black and white of wood-engraving, symbolizing the tension of the contraries, was to be the medium of expression.

I began a thorough preparation, by leading a life of complete asceticism. I did not frequent the cafes; I spoke to no one, except to Anais, 'l'incomparable'; I visited the museums, the churches, the old buildings; and each week I made a pilgrimage to Chartres.

Before attending the Academie Julien, I bought books on Physiology and Anatomy, and studied the development of the embryo, then the growth of the human body, its pliability, its articulation, its movement. Every afternoon I would visit the skeleton-ridden Museum and the Zoo in the Jardin des Plantes.

At this critical juncture, before I could become proficient in my new endeavor, the great Season of Abundance began.

APOCASTASIS

X. *DEUS EST DEMON INVERSUS*

Beneath this conscious will to be The Prophet-Artist, there lurked the intense preoccupation with the problem of rebirth. At the root of my activity lay the desire-germ of a new consciousness, when the dichotomy of the 'I' and the 'me' would be reconciled. For by this time, I had become aware of the division of my personality: the small, actual 'me', and the Taskmaster, the 'I' [I should be inclined to name it the superego, were it not that the Freudian term implies too much a negating function].

To employ the symbology of Blake, the 'me' is Palamabron, who is Pity, and to whom corresponds the Heart [by Swedenborg's law of Correspondence], while the 'I' is Rintrah, who is Wrath, and to whom correspond the Loins. These are the representatives of two Contraries [and 'without Contraries there is no progression']. Between these, Satan, the Spectre, the principle of Negation, thrusts his ugly Head. And these three represent the three classes of humanity.

In the first stages of Blake' classification, we recognize our three protagonists in the three ancient Britons, who were: the Beautiful, the Strong, and the Ugly:

APOCASTASIS

The Strong Man represents the human sublime. The Beautiful represents the human pathetic... The Ugly Man represents the human reason.... The Artist has considered his Strong Man as a receptacle of Wisdom, a sublime energizer.... The Strong Man acts from conscious superiority, and marches on in fearless dependence on the divine decrees, raging with the inspirations of a prophetic mind.*

Later they were transformed into Palamabron the Redeemed, Rintrah the Reprobate and the Redeemer, and Satan the Elect. And in his 'Milton' we can witness the great struggle between these protagonists. It is a replica, on a grandiose scale, of the drama which takes place in us the moment we become aware of the moral problem of the individual in his relation to Society.

But Blake's preoccupation, and my predilection, lay in the Reprobate, for he represents the class of men to which Blake and Blake's Jesus belonged: the Law-Breakers, from whom issue the new emergent values.

Although Blake was clear in his distinction of these protagonists, yet, in the actual drama of their activity, their differences seem to merge together, or interchange and confuse our understanding. But this confusion is justifiable, at least by my experience,

* The Nonesuch Press edition, pages 796-799.

APOCASTASIS

since I found that my 'I' comprised both Rintrah and Satan, and I was impotent in distinguishing their positive and negative differences.

My suffering was the outcome of this impotence. And I was well aware where my task lay. As Blake has admirably said:

Obey thou the Words of the Inspired Man.
All that can be annihilated must be annihilated
That the Children of Jerusalem may be saved from slavery.
There is a Negation, and there is a Contrary;
The Negation must be destroyed to redeem the Contraries.
The Negation is the Spectre, the Reasoning Power in Man..

This annihilation of the Spectre was accomplished at the slow transit of the planet Uranus, the great Awakener, over my 'radical' Sun in its 84-year cycle around the Earth,... and the short but magnificent season of Abundance commenced.

Beginning with a sudden deflation of the 'I' into the 'me', and a complete identification with a Saint Sebastian I had seen at the Louvre ['I became what I beheld'], I baptized myself 'Sansebastian'. For I too, as my torso could show, had the scars of suffering. I too had been 'resurrected'.

Another reason for this identification may have been the knowledge that, to the painters of the Renaissance with their zest for the Open World, the suffering and resurrection of a youth had more significance than the crucifixion of Christ. Saint Sebastian was the Renaissance's substitute for Jesus. And this knowledge appealed to me, at that time preoccupied with the problem of Christianity.

But no sooner was this identification achieved, than I was overwhelmed by the inflation of the 'I'.

As the 'me' took over the legend of Saint Sebastian, so the 'I' took possession of another myth, another legend. Hard as I try to remember, I am unable to trace the circumstances as to when and how I came across the myth of Saint Christopher. An important clue may have been the article in the French occult review, 'Isis', entitled, 'La Legende Alchimique de Saint Christophe'.

This legend, which was also a favorite one with the Renaissance painters and engravers, deals with the Canaanitish giant, Reprobus, Adocimus, or Offero by name, who was resolved to serve only the greatest. First, he served the wealthy and powerful king. But when the monarch betrayed his fear of the 'Evil One', Reprobus abandoned his court, and joined the retinue of Satan. But Satan, too, was to disappoint Reprobus, for, one day as they came to a cross-road, he trembled at the sight of the Cross. This gesture

APOCASTASIS

of cowardice revealed to Reprobus the existence of a greater monarch. Since he was still resolved to serve only the greatest, he turned away from his companions, and set out to offer his services to the powerful one before whose image even Satan had given way to fear....

The effigy of this Saint was usually accompanied by the following inscription: 'Christophori faciem die quacumq' tueris ✦ Illa nempe die morte mala non morieris ✦', which may be rendered: Whoever shall behold the image of Christopher, on that day he shall not die of an unfortunate death. For the sight of his image was supposed to exempt one from the perils of epidemics, earthquakes, fire and flood.[1]

XI. COMMUNIO SANCTORUM

With my interest in Blake's three classes of men, it now seems clear why this legend took possession of me. With a Blakean twist of interpretation, I was able to make the three monarchs conform to a more conceptual pattern. The natural, unregenerate way of life would correspond to the wealthy monarch;

1. I have never been particularly interested in saints. It is possible that there is a connection between the Spanish word for saint, 'san', and the first three letters of my name.

APOCASTASIS

the unconforming and rebellious, to Satan; and the third way to Christ the Crucified:

Positive	Neutre	Negative
Satan	Unregenerate	Christ

I was convinced that these three ways of life had been tried, and had failed. There must certainly be a fourth way, possibly the summation of these three, that would help us to achieve a paradise on this earth.

It was therefore necessary that Reprobus should reappear on earth and offer his services to this new Ideal which was to be, for the most part, a composite of Blakean, Nietzschean and Lawrencian ideals.

Moreover, the passion with which I took over this myth must surely have been due to the cry of distress which escaped from the Sansebastian or the Palamabron in 'me', evoking in the 'I' the peremptory necessity to become the Reprobate and my Redeemer.

Already, on June 6th, 1930, after I had decided to leave for the States, I had written down:

'Cosmical, psychical events within a few weeks... To describe these conflicts that are happening at the moment is for a child to imitate a sunset on a scrap of paper... O when will I know myself? When will I know the motive and the cue for my passions?

APOCASTASIS

Is it when I am old,...and then too late?... O God, it is a new consciousness that I need! A precision, a clarity of thought!... To give new values to things and have them remain so!

'Thank God that the recurrence, the rhythmical repetition of things betray an inner law, a skeleton, a web, an outline to give strength and stability to the flux and reflux of Life!...

'It is this thing Fear that grasps me at the moment of parting, that sucks me before plunging into the Unknown... It is only by a new consciousness that I can bridge this abyss. Not will-power, that is not enough... When will this new consciousness be definitely realized?...

'I feel it emerging like a huge white leviathan rising on a vaporous sea....

'Out of my wreckage I will build again... Like a phoenix I will rise... I must have rebirth. Away with my past education, past prejudices, past fallacies!... Let me feel life, strong, golden life flow through me like beating, pulsating blood!'

The whole Atlantic Ocean lay between that day and the emergence of that huge Leviathan. My Moby Dick was Offero, the Reprobate.

But only in the first few days of the Illumination

had the identification of my whole self with Offero been complete. Later he began to exist by projection. Suddenly, inexplicably, he took on the body of one of the kings of Judah on the west portal of the cathedral at Chartres. And I became the medium in which the drama of the Redeemed and the Redeemer was to be enacted.

Those first few days were ecstatic ones. Never had I experienced such a profound spontaneity, such a smooth play between my afferent and efferent nerves, such a radium-radiation of the mind. How well I understood, then, Blake's representation of Los powerfully striking his anvil!

I then conceived the idea of an illustrated book describing the advent of Offero: his stepping down from his pedestal on the wet evening of November 2nd, All Souls' Day, the day after All Saints' Day; his surprise that the nimbus, given after his death, did not follow him now that he was alive; his first words; his metaphysics of the Contraries; his Adamic outlook; his habit of planting seeds; his creative and warm expression; his affirmations...

And I waited for the redemption: when I would be whole for ever; when I would be actually, completely, irrevocably Offero himself.

APOCASTASIS

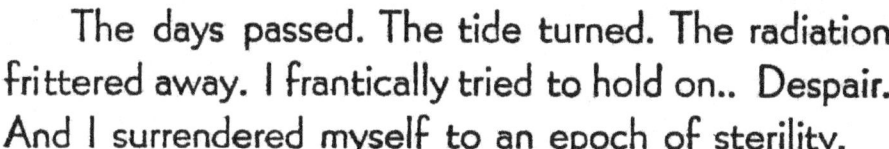

The days passed. The tide turned. The radiation frittered away. I frantically tried to hold on.. Despair. And I surrendered myself to an epoch of sterility.

'Between the idea   Between the conception
And the reality   And the creation
Between the motion   Between the emotion
And the act   And the response
Falls the Shadow   Falls the Shadow'

What was the result of my creative work during that period? Is it possible to believe that it was not worthy, that it was immature? Is it true that this wave of illumination had caught me unprepared? Is it probable that those moments of ecstasy are the moments of fecundation and not of creation?--[What had been the immediate result of that memorable 'l'an de grace 1654, Lundy 23 nov... Depuis environ dix heures et demi du soir jusque environ minuit et demi......'?

I destroyed those notes before I left England. But I have kept one passage which illustrated the psychological drama of the Redeemed and the Redeemer:

And Offero said: Life is a slow curving... All departments of knowledge are slow curvings.... Even

APOCASTASIS

Science... Wait and you will see it burst through the chrysalid of fixed stars and slowly come back to Man and the region of things human. We keep forgetting that Man is the measure of all things.... A bird flies far from home, visits all manner of strange places, and comes back with provender for the young... There is always a going away and a returning.

There is a strange curving we are tracing today. Have you ever noticed the different points of view that the ages have had towards Greece and the primitives? Whenever Nature wants a balance she issues Romanticism. Have you noticed that Romanticism is coming closer and closer to the primitive mind? Isn't this a curving?

Offero came closer, and placed his hand upon me,... and with a voice profound and ageless, said: 'Foolish, pitiful young man! How the old generation and the new generation have harassed you!... I tell you, you are like a snake sloughing his skin, and you are in the throes of agony and despair... Wait. I say, only wait... I am Summer... I am the Sun in full strength... I am your Wisdom... I will help you cast off the old, corrupt slough. And in your joy, in your new strength, you will circumscribe me, holding the tip of your sharp mouth... And we will be one, young

APOCASTASIS

man... You will be I! And I will roll down the hills of the Earth, and the slopes of the skies, making the Music of the Spheres audible and human at last!.... Until the gigantic hand of the West seizes me and draws me down into the Ocean of a new existence!

XII.a. *ASTROLOGIA REDIVIVUS*

I was too impatient to remain long in the subsequent epoch of morbidity. So I turned for help to psychoanalysis again.

This time I was analysed by a 'Breton'. He was a typical Renaissance figure: an homeopath, a champion of the Hippocratic conception of medicine, an astrologer, an occultist, an authority on Paracelsus, and the sponsor of lectures on modern ideas at the Sorbonne.

The surrealist Rene Crevel, who wrote 'L'Esprit contre la raison', and who later, in 1935, was to commit suicide, was once his patient; so was, for a short time, Antaine Artaud, actor, poet and creator of 'le Theatre Cruelle', who was later, in 1939, to become mad.

One would have expected that Dr. X. would out-Jung Jung. As a matter of fact he outdid Freud

APOCASTASIS

in the gargantuan search for sexual derivations.

But it is to him that I owe my deliverance from the stalemate of the moves and countermoves of my Ideal and the Actual.

The means of that deliverance was the introduction of Astrology.

I cannot tell how he came to speak of that dead 'science' [it had aroused in me a fleeting curiosity during my study of Chaucer at college]. It may possibly have been when I remarked that I 'felt' there was a pre-conditioning to the behavior-conditioning-complexes; and that, perchance, Astrology may be the science of that pre-conditioning.

Interested, I asked for a test. A 17-year old disciple and pupil of his 'cast' my horoscope.

I was so convinced with the result that I enthusiastically plunged into a study of the 'science' and its rusty intricacies. Dr. X. and his disciple gave me the preliminary lessons, and a few French textbooks helped me to master the rest.

In the meantime, the analysis was abandoned. It was obvious that this enthusiasm for Astrology, and the determinism that it implied, meant, either the refusal to be cured, or the beginning, in an indirect and paradoxical manner, of the new lease of

of life. Only the future was to tell that the latter alternative was the winner.

For with Astrology I was to achieve an objectivity of outlook I never possessed before. The intense preoccupation with myself disappeared. My superego was reduced to normal proportions. There were no more recriminations, punishment, despair. My 'me' had free scope to immerse itself in the Actual, as it did once before. But the 'I', though tolerant, was vigilant. With the broom of Astrology in one hand, it would chase away chance, and spread the miraculous on the events of my life.

XII.b. *ASTROLOGIA LITERARIA*

Having now mastered the essentials of Astrology, I decided to apply them to a thorough study of Blake.

I soon discovered that the horoscope published in 'Urania' by Blake's friend, John Varley, and republished later in '1001 Nativities', contained a miscalculation. I corrected it. And I corrected as well the position of the Ascendant. I was convinced that, if Varley had known that the planet Neptune [which was not discovered until 1846] had received the attribute of 'Mysticism', he would certainly have tried

APOCASTASIS

to erect the horoscope for 8:15 rather than 7:45 in the evening. For if the rising of this planet is assured, then Neptune becomes the part-ruler of Blake's life.

I here append his horoscope as calculated for London at 8 P.M., November 28, 1757 [the Houses being calculated by the Campanus method]:

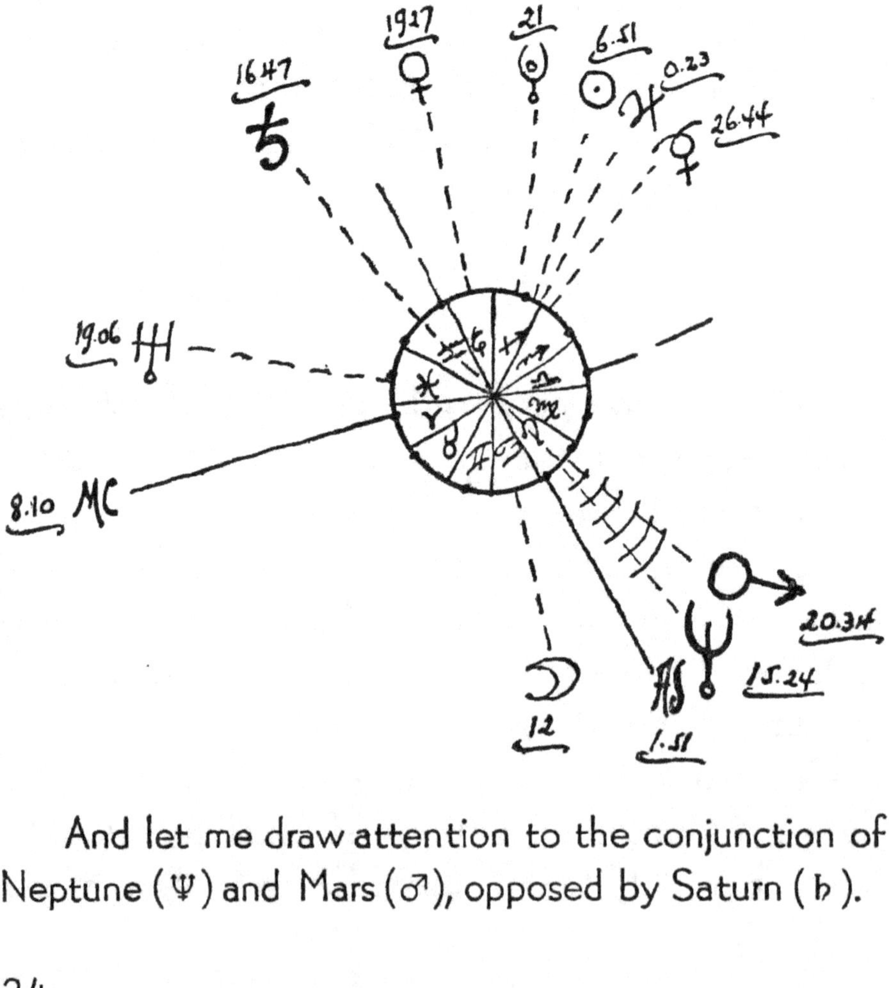

And let me draw attention to the conjunction of Neptune (♆) and Mars (♂), opposed by Saturn (♄).

APOCASTASIS

After savouring the pleasure of my discovery, I began to realize that here was the opportunity to study the conjunction of Mars-Neptune afresh, since here was an example which had never been used by modern astrologers [like Leo, Carter, von Koeckler and Choisnard] for their textbooks.

With enthusiasm I set about collecting the horoscopes of poets, writers, artists and musicians that possessed this conjunction. I was not interested in those of the 'average' person. My field of study was restricted to genius, to the renowned,-- an enigma to Astrology as well as to Psychoanalysis. But I had an imaginative hypothesis which helped to overcome the initial difficulty in dealing with that enigma:

Human beings are like atoms of radium; genius is the disintegrating atom, and in this disintegrating atom the planets operate in all their plenitude.[1]

It would take a book to relate all that I found. It was nothing pragmatically startling. But had I written it, it would have been the crowning [if not the clowning] of literary criticism. For by using this astrological method, I was able to detect the true homogeneous from the heterogeneous. I had here

---

1. 'The man of genius is simply the point of least resistance through which Nature passes into human life.' J. Hinton, The Law-Breaker, 1875.

APOCASTASIS

an infallible guide that would help me to bring together men of genius of different orders, or of different creative fields, and by the 'rapprochement' extend and deepen my understanding of each or all of them.

I did not write it because of my inveterate vice of procrastination, and because this initial search for the conjunction of Mars-Neptune led to the discovery of planetary rhythm, and eventually to cycles and the problem of The Great Year,-- a world I was too eager to explore to turn back and express my gratitude to Blake for the discovery.

In case I am not able in the future to write this token of gratitude, I shall give here the gist of its unwritten content:

No sooner had I placed the horoscopes I had collected in a chronological order, than I perceived that some had their Mars conjunct Neptune and in opposition to Saturn. In the search for other examples of this planetary configuration, which I may have overlooked, I suddenly realized that I was confronted with the problem of time and rhythm. Thus, if it takes Mars two years to come to the conjunction of Neptune, it takes Saturn 36 years to reach the opposition of Neptune, so that quite often [and

APOCASTASIS

this syncopated rhythm can be easily worked out] we find Saturn opposing the conjunction of Mars and Neptune. We find it, for instance, in 1757, when Blake was born. In 1792 [and this time Mars and Neptune are conjoined with the expansive Jupiter], when Blake was in the throes of creation ['Visions of the Daughters of Albion', 1793, 'The Marriage of Heaven and Hell', 1793, 'Songs of Experience', 1794, 'America', 1793], Shelley was born.[1]

Then, when next Saturn begins to oppose Neptune in 1827, Blake dies; and when Mars comes to form the configuration in 1828, William Gilchrist and his wife, the devoted biographers of Blake, are born; and so was Dante Gabriel Rossetti, the poet-painter, and the first enthusiast discoverer of Blake.[2]

Here I had three English poets with a similar planetary configuration. With this configuration in mind, I had only to collect the remarks that critics and biographers had published, wherein they indicated the resemblance between these poets.[3] I entertained the idea that if I were to juxtapose thus, impartially,

1. 1001 Nativities: no. 73.   2. 1001 Nativities: no. 436.

3. Examples: The God of Shelley and Blake, by John Henry Clarke; The Truth about Rossetti, Nineteenth Century Magazine, March, 1883; D. G. Rossetti and William Blake, by R. J. Morse, Englishe Studien.

these quotations, I would not only have proved my thesis, but I would have also helped to extend our understanding of Blake himself.

Nevertheless, this juxtaposition was to be based only on the configuration of Saturn opposing the conjunction of Mars with Neptune. I shall, therefore, try to give an astrological interpretation of this configuration as we find it in Blake's horoscope:

If we turn to Alan Leo's 'How To Judge A Nativity', we will find the following aphorisms:

> Mars is the planet of focussed force and outgoing impulse.
> Saturn is the planet which binds, limits and crystallizes.
> Saturn rules the personal Ego, Mars the animal tendencies.
> Neptune is concerned with psychic evolution.
> Conjunctions stand symbolically for union and synthesis.
> Oppositions signify antagonism, rivalry and duality.
> Mars rising in the As. denotes a strong desire nature.

Moreover, in 'Mars: The War Lord' we find that

> Mars is the force that is ever impelling one outward toward objects of the senses, while the power which is called Saturn is ever drawing one inward for the purpose of reasoning, or rationalising his senses.

We see then that Neptune, which signifies feeling, intuition, vision and higher consciousness, is vitalized by the energy of Mars; and the union of these two planets, passive and positive, is opposed with antagonism and rivalry by Saturn, who binds, limits, reasons or rationalises.

If we keep Blake's 'three classes' in mind, can we not distinguish the Beautiful and Pathetic [Neptune], the Strong [Mars], and the Ugly [Saturn]: Palamabron [the Heart], Rintrah [the Loins], Satan [the Head]: the two Contraries and the Negation?

In 'Milton' the planetary drama between Neptune, Mars and Saturn took poetic form.

But this does not justify us in concluding that this drama, due to a particular planetary configuration, is therefore of small exemplary value for the rest of Mankind. Since we are all subject to these planetary 'functions', whether Mars and Saturn are in a harmonious configuration or not, the contrariety underlying their relationship remains intact. And since it is in the opposition that this contrariety is fully manifested, we must turn to Blake's life and works, if we wish to understand the psychological implication of such a configuration.

Much as the enlightened astrologers affirm the positive value of Saturn [and Dante, did he not place the contemplatives in Saturn's sphere?], they nevertheless agree that he is the taskmaster of our lives. An instructive book could be written revealing the

APOCASTASIS

the slow accumulation of insults that has followed the procession of Saturn down the ages. To enumerate a few: The Greater Malefic, the tyrannical Father, the Miser, the Oppressor. He represents the conservative and reactionary forces, old age, habit, fear, neurasthenia, rationalism, inhibition, the anal complex, etc. He is Blake's Urizen. He is the 'God of this World'.

No wonder, then, that world history comprises epochs of revolt against this god of inhumanity. It is the uprising of Prometheus against Jove; of Los against Urizen; of Rintrah against Satan; of Mars against Saturn; of inspiration against inhibition; of the Id against the superego; of desire against repression!

Today the contemporary revolt is Surrealism. Its intellectual, artistic and moral implications are to be found in that 'black Bible', 'Foyers d'Incendie', by the Greek poet, Nicolas Calas.

If this affiliation between Blake and Surrealism is not arbitrary, we should not be surprised to find in this movement of revolt a Blakean with the Mars-Neptune-Saturn configuration of the year 1900. I have already mentioned him: Rene Crevel [born in Paris, at 5 A.M., on the 10th of August, 1900].

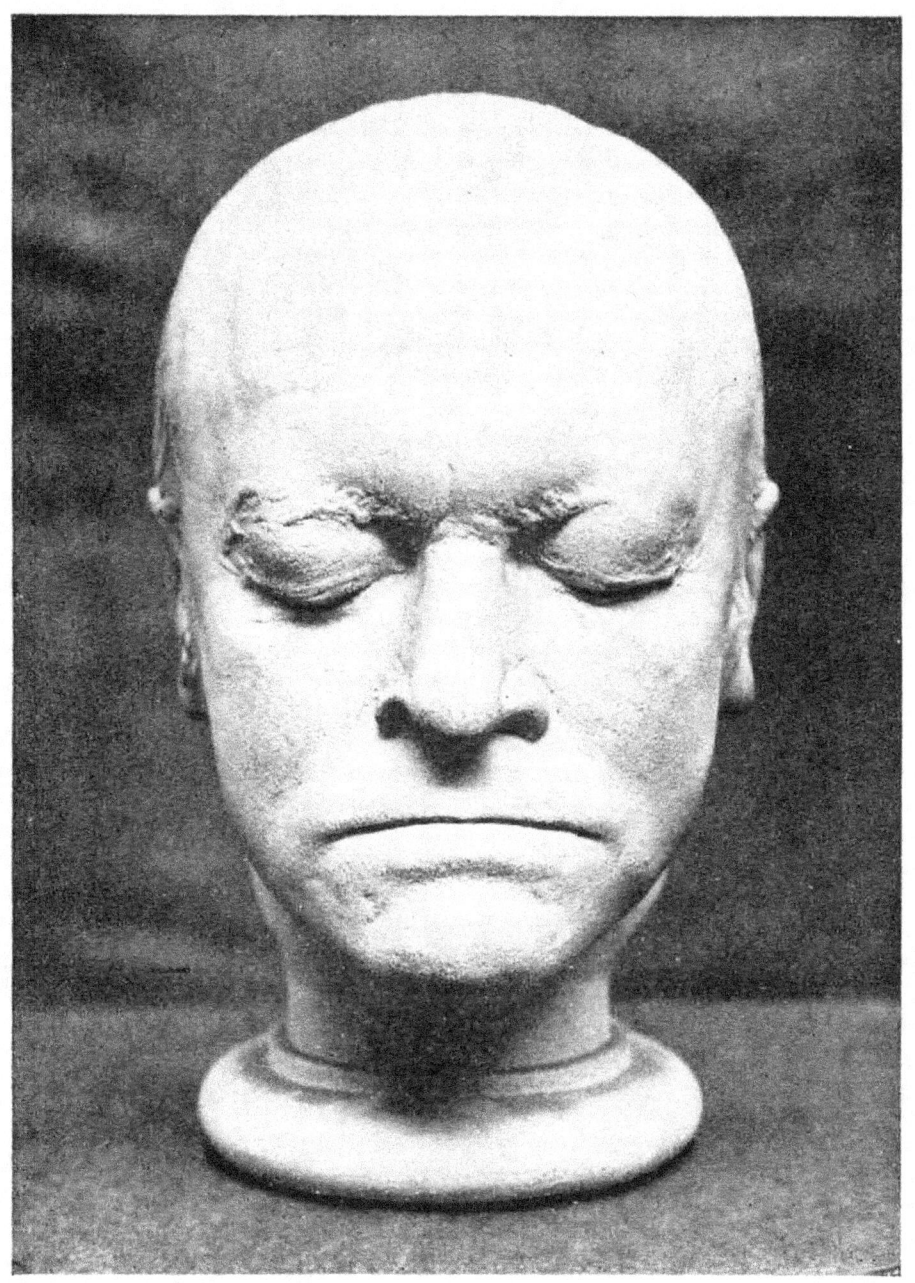

APOCASTASIS

With 'L'Esprit contre la Raison', a Frenchman rises to take up arms against reason, and by this act, France, the France of Louis XIV, expires. The seed that Madame Guyon [born in 1647, at the fruition of the Pluto-Neptune cycle: 1400-1892] had cast: 'It seems that God has chosen me in this century to destroy human reason and install the reign of God's wisdom through the debris of human knowledge....', had now, at the end of the cycle, grown to deadly fruition.

And here ends the gist of the unwritten content of my book on William Blake.

XIII. JOURNEY OF THE MAGI

These astrological researches on William Blake were cut short by the discovery of planetary cycles, which kindled my imagination and goaded me to the task of calculating the conjunctions and oppositions of the superior planets, especially the two furthermost ones yet discovered moving near the ecliptic: Neptune [in 1846] and Pluto [in 1930].

Hand in hand with this search for cycles went my interest in men of genius. These were the objects of my immediate desires. I had to find the cause for

APOCASTASIS

the flux and reflux of life; and I had to satisfy the need for a living Offero [since I had abandoned, after my psychoanalysis, the attempt to be The Seer].

Therefore, when I found that the last conjunction of Pluto-Neptune took place in 1892, I concluded that here ended virtually the epoch which began with the Rennaissance in 1400, and that here, also, began a new epoch. Consequently, somewhere at this time, among the galaxy of men born with the germ of the new epoch, there should be one man who would not abide my questions.

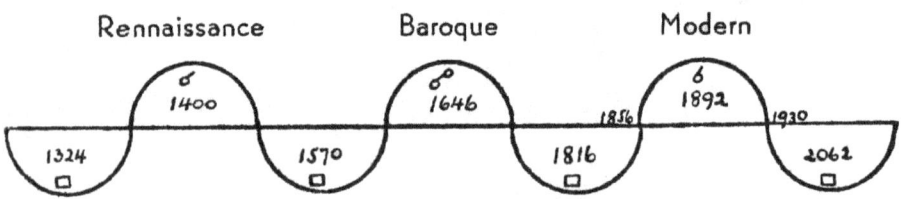

So, as a modern magus, I searched in libraries, in reviews, in conversation with friends, for that new messiah who was already reaching maturity. Sometimes I thought I had found him. But I never felt certain, until the day I read the works of Eric Graham Howe.[1] The unconscious certainty that I

---

1. Born in London, on Feb. 3, 1897, at 3:30 a.m. Of his works I recommend: 'I and Me', 'War Dance', and 'Time and the Child'.

APOCASTASIS

had reached my goal must have been so great, that a few days later [on Feb. 4, 1939, at 11 a.m., to be precise] I wrote my first automatic writing:

My finding Howe has been the last cross erected by Edward the Confessor on the funereal way his dead wife passed.

A few moments of free associations revealed the latent thought of this passage:

History informs us that Edward the Confessor, after his queen's death, erected memorial crosses at the different stages of the funereal procession. A copy of one of them still stands by Charing Cross Station in London.

My name is Edward, and I am the Confessor, for I have 'confessed' to two psychoanalysts.

The station of Charing Cross, like all railway stations, is the terminus of arrival and the starting-point of departure.

Moreover, the immediate association with the word, Charing Cross, was, 'charogne' [Baudelaire's 'Charogne'], a rotting corpse, which called forth the phrase, 'corruptio est generatio'. This last expressed an idea which greatly appealed to me at the time, because I felt that in the phosphorescent corruption

APOCASTASIS

of Andre Breton's Surrealism[1] the seed of the new culture was to be found.

These free associations are enough to indicate clearly that the memorial crosses are the conjunctions of Pluto-Neptune, which take place every 492 years; that the queen-wife is culture or civilisation; and that Graham Howe stands at the end of a culture and the beginning of another.

But I was consciously inclined to believe that, since Howe's metaphysics comprehends the law of Polarity and the law of Cycles, he had transcended, in understanding and awareness, the particular temporal manifestation of the 1892 conjunction. A more fitting example should have been Andre Breton, although, in his awareness of a deeper reality beyond the three dimensional consciousness, he is the equal of Howe.

Therefore, as far as my deepest needs were concerned, Graham Howe was the last cross of the funereal procession. My finding him had ended my search. I had arrived at my destination. The despairing question over decay and death [and the whole

---

[1]. Andre Breton was born in Tinchebrai, Orne, on Feb. 18, 1896, at 10:30 P.M. See his NADJA, 1928, LES VASES COMMUNICANTS, 1932, and L'AMOUR FOU, 1938.

APOCASTASIS

automatic passage is covered over with the pall of that interrogation: the funeral, the crosses...] had been answered.

Now that we have reached the Terminus, we turn and perceive that the psychological problem underlying the whole extent of my experience has been made manifest, and solved. The urge for the pattern of my life, and of Life, has been satisfied.

That 'huge, white Leviathan' became Offero [he who carries the O, the Circle], and Offero, in turn, after my 'I' had abdicated from such a responsible position, became Graham Howe:

'Foolish, pitiful young man! How the old generation and the new generation have harassed you!... I tell you, you are like a snake sloughing his skin, and you are in the throes of agony and despair... Wait. I say, only wait... I am Summer... I am the Sun in full strength... I am your Wisdom... I will help you cast off the old, corrupt slough. And in your joy, in your new strength, you will circumscribe me, holding the tip in your sharp mouth... And we will be one, young man... You will be I! And I will roll down the hills of the Earth, and the slopes of the skies, making the Music of the Spheres audible and human at last!....

APOCASTASIS

Until the gigantic hand of the West seizes me and draws me down into the Ocean of a new existence!

EPILOGUE

Since a quick investigation of the spontaneous passage about Edward the Confessor had revealed to us its latent content, an inquiry into the two legends of the saints, which had so unexpectedly 'seized' my complete attention, should bring fruitful results.

I have already given a tentative, but satisfactory explanation of these 'seizures': the identification of the 'me' with Saint Sebastian, because of my 'sufferings' and the scars on my torso; and the identification of the 'I' with Saint Christopher, due to my interest in the Reprobate.

At this point I need no longer dwell on the 'possession' by Saint Sebastian. Its duration was brief. However, this symbol of Saint Sebastian, being the Renaissance substitute for Christ, was immediately fused with the Child-Christ of Saint Christopher's legend when this, in its turn, engulfed me.

APOCASTASIS

We only need, then, to devote our attention to this last legend.

I have already dwelt on the first part of it where Reprobus seeks the greatest monarch, possessing the fullest, the most complete way of life.

But is this all?

Is it not possible that this legend holds a tremendous meaning commensurate with the upheaval of that 'possession'?

Turn to any illustration of Saint Christopher. Is not the iconography invariably the same: Offero crossing a river with a child on his shoulder?

What is this River? The Stream of Change, of Becoming? The Flux of Heracleitus?

Who is this Child? Is it I? Is it humanity in its infancy?

And what is this gigantic strength that is able to bear the child across Life? What is the secret of its strength?

What was Offero's great prayer?

> As they led him to death, he knelt down and prayed that those who looked upon him... should not suffer from tempest, earthquake or fire... And it was believed that, in consequence of his prayer, those who beheld the figure of St. Christopher were exempt from all perils of earthquake, fire and flood.

APOCASTASIS

These three last words, earthquake, fire & flood: do they not recall to our memory legends of catastrophes wherein civilizations have perished? Do we not remember Plato's 'Timaeus' and 'The Laws', and the Stoics' Great Year with its destruction of the world, now by fire, now by flood?

Is the Law of Cycles the secret?

Is it possible that imbedded in this legend is the immemorial conception of Apocatastasis?[1] And is it possible that the knowledge of this law and the acceptance of this law enable us to wade through the River of Becoming?

These questions are not entirely the consequence of a delirium of interpretation. It seems incontestably certain, as some critics have already observed, that this popular legend is a pagan one, taken over and transformed by Christian hands.[2]

Originally Offero was a dogheaded giant. Anubis, the servant of Osiris, was dogheaded. Anubis was Thoth, the computer of time and the inventor of numbers.

1. See Errata at the end of the book.

2. The result of an inquiry on this line was published in Paris, 1936, by Pierre Saintyves: 'Saint Christophe, successeur d'Anubis, d'Hermes et d'Heracles.'

APOCASTASIS

Offero is the impersonation of Time, of Rhythm, of Cycles.

If the conclusion of my inquiry is valid, that in this legend we find the traces of the ancient concept of The Great Year, is it not possible, then, that this experience, which I have attempted to relate, emerges as one more example to justify Jung's contention that, often, as an individual finds himself in a psychological impasse, his libido regresses until, breaking through his unconscious, it taps an archetypal image of the Collective Unconscious?

But let us be wary of identifying The Great Year with the archetypal image, since this idea of The Great Year is but a conscious elaboration of a far more ancient conception.

Is it possible that the spectacle of the operations of Nature has been deeply engraved in the collective memory of mankind, and that it is this memory which gives birth to the grandiose conceptions of the ancient races: the Vedic 'Rita' and the Chinese 'Tao'?

But is it necessary to postulate this memory? We can just as easily assume that these conceptions are but the biological processes of Nature become articulate in the human mind.

APOCASTASIS

Therefore the individual, who becomes aware of this articulation, undergoes the most profound catharsis that human experience can offer.

ERRATA

Since I have type-set these pages, I am supposed to be responsible for the errata. I query this.

Of course, there are errors due to careless distribution of the type. But these can be easily corrected.

However there are some 'slips' that tax one's vigilance. Others, Puck-like, or fiendishly, elude it. For these, the author-&-typesetter should not be responsible. It's THE OTHER who is to blame.

As far as I know, there is one Puck-like error, and two fiendish 'slips':

Page 12: 'Horrowing'. -- I am rather pleased with this portmanteau word.

Page 30, lines 24 & 25: 'holding the tip of your sharp mouth'. -- On page 45 I was able to improve on it: 'holding the tip in your sharp mouth'. The original text is: 'holding the tip of your tail in your sharp mouth'. -- But I prefer to make no further comments.

Page 42: 'Apocatastasis'. -- When I think how hard I tried to make this a book without errors, it was a stupefying blow, indeed, to realize that the most important word in the text, the most conspicuous word in the book was misspelt. -- It certainly calls for another fifty pages of detection. But time is short, especially to a type-setter.

Anyway, Apocastasis is a better sounding word, and is a more fitting pendant to 'Apocalpyse'. I hope it 'takes', in spite of the saturnian eye of Philology.

As for the lack of French accents, I deeply regret their omission. When I procured the type, I was made to understand that, due to the War, German, French and Spanish accents were no longer to be had.

THE PRINTING OF THIS WORK HAS BEEN LIMITED TO 125 COPIES. IT WAS HANDSET BY THE AUTHOR IN BERNHARD GOTHIC LIGHT, AND PRINTED ON WORTHY CHARTA

Cristofori faciem die quacumque tueris :·
Illa nempe die morte mala non morieris ·:·
Millesimo cccc°
xx° anno :·

ALSO AVAILABLE FROM SKY BLUE PRESS

A Joyous Transformation: The Unexpurgated Diary of Anaïs Nin, 1966-1977 by Anaïs Nin (print, ebook)

The Diary of Others: The Unexpurgated Diary of Anaïs Nin, 1955-1966 by Anaïs Nin (print, ebook)

Trapeze: The Unexpurgated Diary of Anaïs Nin, 1947-1955 by Anaïs Nin (print, ebook)

Mirages: The Unexpurgated Diary of Anaïs Nin, 1939-1947 by Anaïs Nin (print, ebook)

Reunited: The Correspondence of Anaïs and Joaquín Nin 1933-1940 by Anaïs Nin and Joaquín Nin (print, ebook)

Auletris: Erotica by Anaïs Nin (print, ebook, audiobook)

The Quotable Anaïs Nin by Anaïs Nin (two volumes; print, ebook)

The Portable Anaïs Nin by Anaïs Nin, ed. Benjamin Franklin V (print, ebook)

Letters to Lawrence Durrell 1937-1977 by Anaïs Nin (print, ebook)

D. H. Lawrence: An Unprofessional Study by Anaïs Nin (ebook)

House of Incest by Anaïs Nin (ebook)

The Winter of Artifice: 1939 Paris Edition by Anaïs Nin (print, ebook)

Winter of Artifice: American Edition by Anaïs Nin (ebook)

Under a Glass Bell by Anaïs Nin (ebook)

Stella by Anaïs Nin (ebook)

Ladders to Fire by Anaïs Nin (ebook)

Children of the Albatross by Anaïs Nin (ebook)

The Four-Chambered Heart by Anaïs Nin (ebook)

A Spy in the House of Love by Anaïs Nin (ebook)

Seduction of the Minotaur by Anaïs Nin (ebook)

Cities of the Interior by Anaïs Nin (ebook)

Collages by Anaïs Nin (ebook)

The Novel of the Future by Anaïs Nin (ebook)

Anaïs Nin: The Last Days, a Memoir by Barbara Kraft (ebook)

Henry Miller: The Last Days, a Memoir by Barbara Kraft (print, ebook)

Anaïs Nin's Lost World: Paris in Words and Pictures 1924-1939 by Britt Arenander (print, ebook)

Facts Matter: Essays on Issues Regarding Anaïs Nin by Benjamin Franklin V (print, ebook)

Anaïs Nin Character Dictionary and Index to Diary Excerpts by Benjamin Franklin V (print, ebook)

Critical Analysis of Anaïs Nin in Japan, ed. Paul Herron (Print, ebook)

A Café in Space: The Anaïs Nin Literary Journal, Vol. 1 by Anaïs Nin, Janet Fitch, Lynette Felber… (print, ebook)

A Café in Space: The Anaïs Nin Literary Journal, Vol. 2 by Anaïs Nin, Benjamin Franklin V, Masako Meio… (print, ebook)

A Café in Space: The Anaïs Nin Literary Journal, Vol. 3 by Anaïs Nin, Gunther Stuhlmann, Richard Pine, James Clawson… (print, ebook)

A Café in Space: The Anaïs Nin Literary Journal, Vol. 4 by Anaïs Nin, Alan Swallow, John Ferrone, Yuko Yaguchi… (print, ebook)

A Café in Space: The Anaïs Nin Literary Journal, Vol. 5 by Anaïs Nin, Duane Schneider, Sarah Burghauser… (print, ebook)

A Café in Space: The Anaïs Nin Literary Journal, Vol. 6 by Anaïs Nin, Joaquín Nin, Tristine Rainer, Christie Logan… (print, ebook)

A Café in Space: The Anaïs Nin Literary Journal, Vol. 7 by Anaïs Nin, John Ferrone, Kim Krizan, Tristine Rainer…

A Café in Space: The Anaïs Nin Literary Journal, Vol. 8 by Anaïs Nin, Benjamin Franklin V, Anita Jarczok, Kim Krizan… (print, ebook)

A Café in Space: The Anaïs Nin Literary Journal, Vol. 9 by Anaïs Nin, Anita Jarczok, Joel Enos… (print, ebook)

A Café in Space: The Anaïs Nin Literary Journal, Vol. 10 by Anaïs Nin, Benjamin Franklin V, Kim Krizan, William Claire, Erin Dunbar… (print, ebook)

A Café in Space: The Anaïs Nin Literary Journal, Vol. 11 by Anaïs Nin, Henry Miller, Alfred Perlès, John Tytell… (print, ebook)

A Café in Space: The Anaïs Nin Literary Journal, Vol. 12 by Anaïs Nin, Kim Krizan, Benjamin Franklin V… (print, ebook)

A Café in Space: The Anaïs Nin Literary Journal, Vol. 13 by Anaïs Nin, Barbara Kraft, Danica Davidson… (print, ebook)

A Café in Space: The Anaïs Nin Literary Journal, Vol. 14 by Anaïs Nin, Jessica Gilbey, Joaquín Nin-Culmell… (print, ebook)

A Café in Space: The Anaïs Nin Literary Journal, Vol. 15 by Anaïs Nin, Rupert Pole, Steven Reigns… (print, ebook)

A Café in Space: The Anaïs Nin Literary Journal, Anthology 2003-2018 (print, ebook)

ANAIS: An International Journal, Anthology 1983-2001 (print, ebook)

ANAIS: An International Journal, Vols. 1-19, 1983-2001 (ebooks)

www.ingramcontent.com/pod-product-compliance
Lightning Source LLC
Chambersburg PA
CBHW081356130526
44581CB00012B/97